LOCOMOTION

For railway enthusiasts past and present

Crumps Barn Studio
Syde, Cheltenham GL53 9PN
www.crumpsbarnstudio.co.uk

Cover design and illustrations by Lorna Gray
Photographs reproduced with the kind permission of the copyright holder. Our grateful thanks to the community of photographers who provided images

Typeset in Adobe Garamond Pro

All our books are printed on responsibly sourced paper from a mixture of managed woodlands and recycled material. Printed in the UK by CMP, Poole.

ISBN 978-1-915067-75-3

LOCOMOTION

A POCKETBOOK CELEBRATION OF
200 YEARS OF MODERN RAIL TRAVEL
IN MEMOIR, HISTORY AND VERSE

DIANA ALEXANDER • JAMES ARMSTRONG • CAROLINE BRADLEY
MYLES CUTLER • BRONWEN R. EVANS • ASMA KHAN
REBECCA MCDOWALL • SALLY MILLS • JEFF NICHOLLS
F. DAVID POTTER • GRAHAM POWELL • MARGARET ROYALL
STUART SAMUEL • ANNE SWAN • TIM TAYLOR • JULIE WILTSHIRE

Crumps Barn Studio

Locomotion No.1 was built by father and son engineering duo George and Robert Stephenson at the pioneering locomotive manufacturing company Robert Stephenson & Co. in Newcastle. Robert would later design and build Rocket, winning the Rainhill Trials in 1829

FOREWORD

On 27 September 1825, the steam engine known as *Locomotion No.1* hauled the first passenger train on the opening day of the Stockton & Darlington Railway. This was the first locomotive to run on a public railway and it paved the way for modern rail travel.

In this era of regular (relatively!) timetabled services where a person might start in one corner of the UK and travel entirely by train to another country, the significance of this event is profound, and yet also perhaps frequently understated.

It's easy to forget that at the time of its inception, modern rail travel opened up mobility for the lower classes. It empowered many Victorian women to travel in relative independence, and improved communications and trade with industrialised areas. A journey which might have taken days by horse-drawn carriage suddenly took only a few hours by train.

Even now, 200 years later, train travel still carries the same allure. My own relationship with rail travel has been lifelong. My father's photographs of steam trains were the stuff of family legend. I was convinced I met the real Father Christmas on a Santa Special, and I puffed across the Yorkshire moors with my grandmother whose own childhood memories of main line steam were rather less romantic, since all she ever seemed to do was get soot in her eye. I've explored many of Britain's heritage lines with my stepfather, and main line rail has always played its part too, particularly now, when my mother lures me down to London from my Gloucestershire home.

The original steam locomotive that made all this possible was built by *Robert Stephenson & Co.* in Newcastle. It was the creation of engineers George Stephenson and his son Robert.

Born in 1781, George initially worked as a brakesman, controlling the winding gear for a Northumberland colliery. He studied reading, writing and arithmetic at night school and rose to the rank of enginewright, ultimately designing and building a number of travelling steam engines for hauling coal. He also helped to further the progress being made in rail design, and his Stephenson Gauge became the basis for the modern standard gauge used on main line railways.

His son, Robert, was born in 1803. Determined to correct the hardships of his own childhood, George sent Robert to school and at the age of sixteen, Robert was apprenticed at the same Northumberland colliery which employed his father in Killingworth. He was particularly skilled in surveying.

When George was commissioned to construct a line for

the Stockton & Darlington Railway, the committee's plan originally incorporated horse-drawn wagons. George quickly encouraged them to amend these plans.

With approval to survey the route and design and build a steam locomotive fit for the task, Robert left his apprenticeship and the father and son team established a new company to manufacture locomotives with the younger man as the manager.

Robert Stephenson & Co. was the first company in the world to be established solely for the purpose of building railway engines.

Originally named *Active*, the engine known as *Locomotion No.1* made the inaugural journey on the new line. Its load was impressive. More than 300 people were on board – with the dignitaries travelling in a specially built carriage and the rest in twenty or so additional wagons, along with a further eleven wagons of coal. The era of public rail travel had begun.

When I first began planning this new collection of short histories, memoir and poetry, I imagined the authors involved would reveal many different aspects of rail travel. But I was not prepared for the real power which populates these pages.

For some of the authors, it is the history of the railway that is important, whether their own or the stories of the people who built them long ago. For others, rail travel speaks of the power to connect people and new landscapes in the present day.

But every line has one element in common. Whether celebrating the mighty engines as they work, or tracing the shape of the people who made it possible, this book is a journey – and that, after all, is what rail travel is all about.

Lorna Brookes
Crumps Barn Studio

Blackpool Central Station, with Blackpool Tower behind (Ben Brooksbank)

BUCKET AND SPADE SPECIAL

STUART SAMUEL

"Wait and see what the weather's like – now get off to sleep."

Fat chance – I was too excited. Blackpool was on my mind.

Morning eventually dawned. An early call, "Get up, we're going."

Unlike being summoned for school, I needed no second shout.

Soon Mother, Father and I were following the others down Horton Street, attracted by an advert in last Thursday's *Courier and Guardian* for a 7s 6d excursion. Organised in the days before accountants ran the railways, they gave the non-corridor coaches, usually dormant for most of the year in Low Moor carriage sidings, a little trip out, and used engines, whose only requirement was to be able to move, from Low Moor shed. These were usually 'no heat' freight engines, meaning that there was no heating in the carriages, but that was of little concern, as you hoped the weather would be warm.

Your carriage consisted of a six a-side bench stuffed with horsehair, which was very prickly to the legs of the unwary child wearing short trousers, and a window you let down with a leather strap. And you were hauled by a friendly, willing crew, for they were being paid to venture to Blackpool, berth the engine in the shed, toast their toes on the beach and take an occasional paddle, before working back. If the weather was sunny and the day promising, more coaches would be added to the train, regardless of the engine's ability to haul them, for, who cared? They'd get there somehow. Everyone was in a holiday mood.

On entering Halifax station, you queued to buy your tickets from Alan, the friendly clerk. You never thought to book in advance – it could be raining, and there was only one thing worse than being at home in the rain, and that was being in Blackpool in the rain - and then proceeded via Albert or Kenneth the ticket collectors, whose clippers were warm with use at this stage, to platform 2. There you found a space near the platform edge, as the crowds usually thronged the whole length of the platform, and awaited your train's arrival from Bradford, having called at Low Moor and

Lightcliffe. The value of non-corridor stock was there was usually a door adjacent when the train pulled in, and you clambered aboard and awaited the guard's whistle for the beginning of your adventure.

But you got no further than Sowerby Bridge, five or six miles or so down the line, before stopping again. And then again at Luddenden Foot and a little further along at Mytholmroyd. Both these stations had short platforms, so the train had to stop and then draw-up so the rear coaches would be platformed. This all took time, but, who cared? You were having a day out. Incidentally, in those days before Health and Safety, locking doors and selective door opening, the railway was remarkably relaxed about stopping with carriages not being in the platform. If you opened the door and jumped out without checking a platform was present, and broke your leg, it was your own fault, and no use running to the company for compensation.

The final station stop was usually Hebden Bridge. Now, at long last, you could start making inroads into the journey. But only to get a few miles further on to Hall Royd Junction, where you left the Calder Valley for the climb over Copy Pit summit into Lancashire. Here you awaited the banking engine to bang into you from the rear, for this engine crew were not enjoying a day out in Blackpool but condemned to spend the whole shift going up the hill to Copy Pit and back. After much whistling, the whole ensemble slowly set off, the noise deafening as it slowly wound round houses and mills around the back of Todmorden. Pity the man sitting in his garden on a bank holiday Monday – your train was only one of several to go clambering by. The banker dropped-off at Copy Pit summit, leaving you to amble down through Burnley and over the rooftops of industrial Accrington and Blackburn, admiring the many terraces of houses and their forests of television aerials. 'Amble' here is the operative word for because of congestion due to a number of trains ahead, you usually moved from signal to signal, and the reason why one of Low Moor's speedier express engines was not strictly necessary became painfully apparent.

At this stage, the prepared unscrewed the flasks of tea or coffee they had brought with them – buffet cars were unheard of on these jaunts, and if there were one somewhere in the train, you couldn't get through to it. Mother peeled the orange she'd brought to share with me, and Father had fallen asleep, his paper only partly read in his lap.

For me, the railway enthusiast, we were approaching the most interesting part of the journey – the passage though Preston. Here, your line from East Lancashire crossed the West Coast Main line from London Euston to Glasgow, to branch out to Blackpool and The Fylde Peninsula. You rarely got through without being stopped at least once, for being an excursion you were of the lowest priority and were simply slipped in between the more important *Royal Scot* or *Jubilee* hauled expresses. Then, passing the fine spire of St Walburge's church on our right, you began to pick-up speed along the four-tracked section to Kirkham and Wesham, where you could either turn left to Blackpool via St Annes and the coast, go right to Poulton-le-Fylde, Blackpool North and Fleetwood, or straight on via the direct Marton line to Blackpool South and Central. On this final section, when excitement was mounting, Mother usually set me the task of looking out for the first sightings of Blackpool Tower, not that she was particularly keen to see it, but it kept me quiet for a while.

Eventually, after stopping at Blackpool South, where no one seemed to alight, the train wearily pulled up past the carriage sidings, locomotive shed and football ground into Blackpool Central, a massive station with its platforms supplemented by six or so excursion platforms to the right. Here you checked your legs were still working properly and headed to the first port of call – the public conveniences. Well, you had endured almost three hours on that non-corridor train. Then it was into Woolworth's café opposite for fish and chips, then a play on the beach to justify the bucket and spade and finally a trip on the tram to the Pleasure Beach and massive funfair, for the main purpose of the day – to part your parents from their savings as quickly and efficiently as possible.

Tired and exhausted you returned to your train about 6pm or 6:30pm, hoping for a speedier run back. A triumph of hope over experience if there ever was one. For everyone was going back. The lines were just as congested. The bankers over Copy Pit less enthusiastic, the drawing-up at Mytholmroyd and Luddenden Foot was tedious rather than interesting. Horton Street seemed more of a drag than usual, and you fretted if you'd make it in time for your last bus home.

Never again you promised yourself … Until the next Bank Holiday.

Such were the rigours of 1960s train travel.

- *Today you can do the journey in half the time and there's an hourly service;*
- *Copy Pit holds no terrors for modern, high-powered diesel trains. Bankers are just a memory;*
- *There is little trace of neither the shed nor the carriage sidings at Low Moor;*
- *Trains use Blackpool North – the track bed of the Marton line is under the M55;*
- *Luddenden Foot station is closed.*

STATION TO STATION

GRAHAM POWELL

I'm in the Quiet Carriage once again:
With a tribe of assorted travellers
Pressed tight – bracing ourselves – waiting for
The lurch to slew our papers onto
Each other's laps or tip our coffee
Into pools across the tabletops.

We have our books, our laptops and our phones
For company so we're not alone –
Always eager for some distraction;
For someone to say they're missing us
And that they're waiting at the station
Or that the meeting has been cancelled after all.

You text to say that *Sirens* got you beat
While I am rocking gently to the sound
Of nothing but the rattle on the line;
Trees pass, I see my reflection close at hand
And then a farmer walking to a gate
Bucket in hand, quite unaware of me.

My neighbour asked me – hours ago –
If I was going far; he's been restless ever since
Unable to settle, half–asleep, fretful
At his silent phone. He moves when he can.
Finds a seat with room to stretch himself out,
Arrange his coat and bag and go to sleep.

Across the aisle a student's reading Pope.
Making notes in the margin, underlining,
Highlighting the key words with a pen that's
Moving way beyond control – he gives up
And looks out of the window: wet orchards,
A lock gate being turned, another halt.

There are new arrivals now, holding hands –
Looping fingers, tucking into each other,
Settling down to watch some romantic film,
Sharing their headphones – each has one earpiece
Snugly plugged into adjacent holes
Well before the titles start to roll.

We're all just killing time until we're there –
Stroking chins, curling strands of hair,
Doodling at a crossword and smiling when
A word falls out or we hear that song again
Or see those lovers kissing in the rain
Or find – like me – this final stop to end a line.

The West Highland Railway between Fort William and Mallaig reflected in Loch Eil (Jeff Nicholls)

THE WEST HIGHLAND RAILWAY

ANNE SWAN

The West Highland Railway, which runs just north of Helensburgh from Craigendoran to Fort William, some 100 miles, was started in 1889 and completed in 1894. The Extension to Mallaig, a distance of 40 miles, also took five years; 1897 to 1902. It is one of the most scenic rail journeys in the British Isles, for the train travels where no cars or people can go, through isolated and desolate areas as well as by lochs, along glens, past hills and mountains – all places of great beauty. Fans of Harry Potter will have seen the film where the steam engine chugs across the Glenfinnan Viaduct.

In 1965 John Thomas published his book *The West Highland Railway*, which has been reprinted and updated and gives a fascinating account of the difficulties faced, events that occurred and the lives of some of those involved in its building. Other sources give slightly different dates, distances and spellings, McEwen/Macewen, so it seems best to stay with John Thomas; any errors or inaccuracies are mine and much regretted.

Seven men set out to survey Rannoch Moor in winter in 1889, some dressed as for the office (two carrying umbrellas). Three engineers, a solicitor, two factors (estate managers) and a contractor all got lost. The weather was dire, the night descended on them, they failed to meet up with their guide, the boat used to cross a loch was leaky and they had to use their boots to bail it out – everything that could go wrong did.

Keepers heard them and took them to the lodge where they expected to spend the night. But no message had been received, so there were no beds or food for them. The lodge keeper produced blankets and knocked up a meal. The lodge was cold and damp. Next morning the party set off again, although it was already raining, eventually heavily; but after seven miles and having had their lunch, they should have met the local landowner. Instead, a keeper came and suggested they go to the lodge for the night as there were only three hours of daylight left. But it was eight miles to the nearest cottage – which didn't sound far – and they decided to push on. Unfortunately, the terrain was exceedingly difficult and when it was too late to turn back they realised they would have to keep going and hope to reach the cottage. While they may have had matches, they did not have torches as we know them. Walking was not easy and trying to keep to dry areas was nearly impossible, particularly as it grew dark; they found themselves stumbling about, tripping over and into bogs, peat and brushwood. They became exhausted.

A sixty-year-old then collapsed, so two of the engineers stayed with him. His umbrella proved useful as a tent. Then two more gave up, could go no further and sought refuge behind a big boulder. The other umbrella provided some shelter. It was now night. One of the two men still standing was a civil engineer who'd walked the moor once – but in daylight and good weather. He set off to try and find the cottage. In the dark, he walked into a fence, knocked himself out for four hours; but realised on coming round that the fence could lead to the cottage. It did. The shepherds there put him to bed and then went to rescue the others. Having found them, after some shouting, they half dragged and carried the sixty-year-old to a hut two miles away, where the five exhausted men spent the night drying out their clothes by an ancient stove that their rescuers got going using old peat turfs. Next day the shepherds came back with food and then, with the ill man on a cart, the survey party – only the contractor missing – started on the last eight miles to Inveroran.

The contractor had also set out in the dark in dreadful weather to get help, but lost his way and spent ten hours on his own on the moor. He tripped and fell so often that at one point he took his clothes off to wring them out before putting them on again. He must have looked a sight when daylight came and he finally found the cottage; there he was put to bed for a few hours. He then joined the rest of the party, and all made their way to Inveroran for the night.

The next day there was a blizzard blowing over the West Highlands.

John Thomas describes vividly the life of the navvies who worked in those isolated areas. Everything they needed to live – wood to build their huts, forty men to a hut, their food etc; and the equipment and material needed to build railways, all this had to be transported from civilization to the wilds. Camps were set up along the line at various distances, each camp having a first aid post with permanent nurses. There was also a telephone to headquarters in Helensburgh so progress – or lack of it – could be monitored. The navvies blasted rock, dug earth and smashed up stone; but when it came to Rannoch Moor they had to use earth, turf, brushwood, ash and spoil to create 'cushions' on which to rest the rails. But this all sank as soon as it was laid on the boggy ground and more ash and soil had to be transported by rail from the south until, eventually, summer and dry weather enabled completion of that last section.

The contractor on the survey of the moor was Robert McAlpine. His early years, schooling and work are already a book in their own right. Coal mining was the main (possibly only) employment in Lanarkshire where he was born in 1847. He skipped school, leaving at ten to be a *trapper*, then a *drawer* in the coalmines, and began on the coal face at the age of fifteen. An apprentice bricklayer at seventeen, he continued working in the mines in order to earn more; he also went to night school as he realised he needed to be able to 'read and figure'. At twenty he married and the following year set up on his own as a jobbing builder.

Over the years he built up his business – there were many challenges – and in the 1870s he became known as Concrete Bob, due to his experiments with it. Concrete had not previously been used much in Scotland. Now, in 1897 with his four sons in the business, he was building the Fort William to Mallaig Extension.

McAlpine's son Robert Jr (aged 26) was put in charge of the Extension, with younger brother Malcolm (aged 19) as his assistant. Four camps were established along the line, the largest at Lochailort housing some two thousand navvies. The schoolhouse was turned into a hospital with eight beds, two nurses and a resident doctor. McAlpine was the first person to do this for his work force – he felt strongly that they shouldn't be too far from a station. But in these isolated and difficult places getting and retaining good labour, craft or tradesmen was a great problem, many only coming for a short time – before heading south where life was easier.

Then tragedy struck – Malcolm was overseeing the blasting in a cutting and despite being 1200 feet from the explosion, he was hit by a lump of stone and many bits of rock. His side was crushed, pelvis broken and there was internal bleeding. He was taken to the hospital at Lochailort where the doctor did what he could. A message was sent to his father at Helensburgh saying he'd been badly injured and was not expected to live.

It was evening, but McAlpine went straight to the Glasgow home of Prof. William McEwan, one of the finest surgeons in Scotland. He was out, not due back for some time and could hardly be expected to drop everything, leave his patients etc and go the wilds of Scotland that night. But McAlpine was desperate and begged and urged and eventually persuaded McEwen, saying he had a 'special' (engine with a wagon) ready to take them up the existing line to Fort William. The request for a special had been met by 'not possible, out of the question, no one to work the switches etc', but McAlpine said he would take responsibility for the journey. He also ordered the foreman of the break-down gang to prepare to come too.

It was 10:30 that night when the special left. The break-down foreman was told to be in the engine cab, get down at every switch, unlock it, see the train through, re-lock the switch and climb back into the cab. All totally illegal, but the situation was desperate. They reached Fort William at about five in the morning, woke up the stable hands and hired a coachman and horses to take them the next thirty miles over rough 'road' to Lochailort. On arrival, McEwen went straight to the hospital and operated on Malcolm – the operating theatre will have been very different from that in Glasgow.

McAlpine feared his son would die if the professor left. For four days McEwen sat by the patient – then decided they needed to get to Glasgow if they wanted to save him. Any jolts would be fatal, so horse and carriage could not be used. McEwen selected eight navvies and trained them for some days till satisfied they were proficient in carrying a stretcher. They then carried the patient along the road to a loch. The men went into the water and lowered the stretcher into a boat; this was towed by another boat across the loch. The men went into the water again to lift the stretcher out and carried it seven miles to Glenfinnan. The hotel doors were too narrow, so a window was knocked out and the stretcher carefully passed through and placed on a large table for the night. At dawn they carried it seven miles to a section of railway where an engine and wagon awaited. With a long coupling the navvies sat opposite each other on the engine buffer beam and the wagon frame, legs touching to absorb every jolt and protect the patient in the wagon. On slowly to Locheilhead, where they transferred to a steamer that sailed to Banavie where a special ambulance train took them to Glasgow. The navvies then carried the stretcher through the streets to McEwen's nursing home.

What an incredible story. Malcolm lived to be ninety. McEwen became a pioneer in brain and bone surgery. While the railway remains a wonderful journey.

Glenfinnan Viaduct (Jeff Nicholls)

ISFIELD

TIM TAYLOR

Isfield – a little station on the old London Brighton and
        South Coast Railway Line 1889
between Uckfield and Lewes in Sussex,
a quiet little village preserved in time.
The railway is now called 'The Lavender Line',
named after the coal merchant who had his business there,
(not after the fragrant smell that one associates
        with lavender).

Immortalised in tiles at Victoria Station
and in lines written by Turle in 1934
"If I were a very rich man, I think I would buy Isfield
        Station from Southern Railway ..."
a country station – preserving the manners and
        courtesies of old
and the fascination with steam railways
for young and old alike.

There is a comradeship, courtesy, humour and
purposefulness
about the men who now work there
like the same men of steel who built the line
all those years ago.

Today's men work to restore the track and engines
to bring back the age of steam
and realise a lifelong, cherished dream.

Time is of no consequence
because the work will go on regardless
into the coming years –
the work will never finish,
because that is what they do –
if it ever finished there would be no purpose,
nothing to look forward to.

There'd be no more sweeping through the heavy gates,
no more cheery greeting of your mates;
no more use of the restaurant carriage
for the Reception after your marriage –
no more guests, no more Groom
anxiously pacing the waiting room
(until he sees the tables spread
with fine white linen cloths, gleaming glasses
and menus waiting to be read).

The engine arrives at the platform
with much steam and hissing –
gleaming brass and copper pipes,
greased axle bearings,
shiny piston rods and fairings.
The evocative smells of the coal fire
and the melting steam
drift across the track
and we are all taken aback
by the sudden piercing blast of the steam whistle
which makes us jump and
causes little boys' mouths to drop open
– their fathers smile
and their grandfathers suddenly remember
the day they steamed their last steam mile.

Railways –
byways and byelaws
'sleepers'
and 'chairs'
antique wagons
and rotting wood
rusting iron
peeling paint and dripping gutters –
this is The Lavender Line.

Steel lines running up the track
a mile to Worth Halt and then back
ending at the buffers.
You pass men, busy in the sheds,
cleaning and painting the coaches
a lovely glossy green.

In Cinders Café
(the old Ticket Hall)
there is a wood burner,
(a warming fireside on cold winter days)
gas lights and lanterns,
enamelled signs from old railways,
brass clocks and signals –
and outside on the platform,
lamp posts and
red fire buckets full of sand.

The men have a full English breakfast
before
shunting the engines
into the siding
and hunting for tools
which are hiding.

Dreamy days
Steamy days
Oily days
Wet days
Sunny days
Happy days
Nostalgic days
Innocent days.

Isfield Station on the Lavender Line in East Sussex (Tim Taylor)

DEPARTURE IS SUCH SWEET SORROW

A recurring scene at Newark Northgate Station

MARGARET ROYALL

The train pulls out.
You wave to me from the window. Such partings
    are difficult.

Already your focus is shifting,
thinking of home and work, your partner waiting for you.

My car is slow to pull out,
as though my heart is still attached to the rear coach

by invisible puppet strings that
someone forgot to untie before the whistle blew.

My focus won't easily shift.
I'm still wrapped in your hug, texting *lovely seeing you.*

And you are texting your partner
Just leaving, call you from the station before home.

And I'm holding back the tears,
biting my lip so hard, determined not to let sadness win,

calculating how long it may be
before your next call asking to come up and stay again.

I'd counted the days to your visit,
yet even before you arrived I was grieving your departure.

But I don't want to suffocate you,
so I try to stay calm, not fuss, not be an embarrassment,

though inside I'm dying slowly,
and I ask myself *is this what a mother's love is all about?*

Is he sad at leaving me?
You are grown up now, fled the nest, lead your own life.

But in there somewhere I hope
you feel a pull of strings too, but then untie them, fly free,
as you should.

RULE OF THUMB

JAMES ARMSTRONG

Look there, through this morning's thin mist. Discern that heightened passage dividing pale fields from a tangle of rising woodland: civilising ruled marker on the unruly tumble of country. It implies a beginning and terminus; intention to inject certainty and timetable into abundant country and life.

The Victorian earthen mound remains: a green stripe capping meadowland, still. On it our former railway line, now RIP. My OS map tells it as 'disused' – not 'unused' – actively taken from us, the travelling public; should read 'disabused' of our treasured rail-route. Thanks.

Can you see there, horses hauling in cartloads of soil; tough, rough navvies, heaping up a railway embankment? Do you hear the endless clattering of them shovelling clinker on, the gang-master bellow, 'I want it spread level like a blanket, you rabble.'

I'm watching now, some shirtless workers, their arms

levering, with long staves, steel rails into place; their sledgehammer arcs piling down on sleeper bolts, hammering clamps home.

Study their steel-toed boots, planted like tree boles; their dust-caked trousers and once-white vests. Examine this one's sun-browned face, grimy upper torso, blue-scarred forearms. Look closer at another: hands hard as the hickory pickaxe shaft he's wielded since his cut-short youth. His fingers twisted into shapes unnatural by endless graft. His clothes reek of pipe smoke, earth and stale sweat.

Down the track, I glimpse brickies working off high timber scaffold. Focus, and you can just make out an image of the bridge and red tunnel-arch they're building, course by course, over an intricate timber former. They're clarting the mortar with a passion. Below them are labourer teams carting sand and cement from close-by Budleigh-on-Sea, brought upriver to Otterton leat. Others are mixing mortar while hod-carriers shoulder packs of bricks up steep ladders.

Observe nearer, hawk-eyed surveyors and assistants checking levels, heights, distances; testing mortar-mix proportions, counting bricks used and yet to be. Watch overseers confirming brick-courses per hour, sleepers laid per day, yards of rail-line per week.

'By rule of thumb, you should be building eight courses a day. Your squad are falling behind.'

'There's much brick-cuttin'. They're working flat out, Guv.'

'Easy replaced if they don't want to work.'

Witness the mini-society that sprouts among fields for a time – stables, tent-sites, latrines, wash-shops and wooden dormitories: all kept close to the site to reduce travel time, maximise man hours. Eventually, the caravan migrates inland to virgin soil as laid track gets signed off and new embankment marches on.

Peek now, inside one dormitory, through smoke from clay pipes; discover the men greedily refuelling with food; 'Not bully-beef again?'

'Aye, but diff'rent dressed.' The Yorkshireman cook says. 'Two choices: tek it or leave it.'

They while away rest-time drinking, gaming – dominoes, shove-h'apenny. Some play at cards or the Put 'n Take spinner; the foolhardy gamble with cash, 'You've just lost a day's wedge.' Or the prudent, with spent matchsticks and wives' voices at their ears; 'Don't be a'squanderin' what you've grafted for with hard chisel and muscle. Scrimp all you can while there.'

Harmonica or banjo sounds might float behind their telling of jokes or bickering over tall boasts. 'Oi've laid eight 'undred bricks today.'

'But for us hoddin' and a'mixing grog for ya, you'd niver

'ave laid a dozin.'

A few are perhaps lying in their bunks, chasing down some urgent shut-eye before unwelcome dawn rushes to meet them all too soon.

Now feel, as they must have, their satisfaction and pride of work when looking at white steel rails narrowing away, behind them; turning with regret to look ahead, envisioning the track yet to be laid before they could afford to return home to Ireland or Italy with a roll of notes in their calloused palms or dusty pockets.

The Minister of Tarmac – paid half a million a year in 1965 money – wielded his axe to the railways to 'save eighteen million pounds a year'; named it 'no longer cost-effective'. Was that fact, sixty years ago? They never proved it. Promised bus-services (poor compensation) arrived much later. But public money invested in construction, land compulsorily grabbed, the bodies of countless navvies degraded while gouging out the line *and* seventy year's worthy use could have been given more slack, surely?

Can you picture trains steaming across the land, packed with servicemen, bound for training exercises on Woodbury Common, or to the Somme, or to Dunkirk?

'By jingo, we'll batter the Boche.'

'We'll send Hitler running for cover.'

Later, hordes of holidaymakers in short sleeves, beach-bound. 'Who'll be first in the sea?'

'Me, me, me.' All three children shriek.

There puffs a train; passengers waving at dog walkers in fields or work-bound folks paused at a level crossing.

Earlier, one February evening, mid-1890s, councillors are sitting round a table. Chairman Merryfield speaks: 'To construct a railway link between Exmouth town and Sidmouth with stations at Littleham, Budleigh, Tipton and Ottery. I trust everyone has studied the proposal carefully? Submission and objections must be considered then our recommendation made to full council. By rule of thumb, we should be done by ten.'

Councillor Hocking opens with, 'Many see the undoubted advantages rail travel has brought, these past forty odd years.'

'There have been objections to noise disturbance and unsightly rail-sidings,' reminds Councillor Barton, 'spoiling our countryside.'

'Far fewer than anticipated,' says Councillor Venn.

'True. Increasing visitors to our seaside towns can only help sustain them,' says Councillor Harris, owner of two hotels. 'Out of season, too, with our mild Devon climate.'

Chairman: 'Thankfully, the estate landowner agrees to lease the largest area of route avoiding much compulsory purchase.'

'These places already suffer decline and deprivation.' Councillor Hocking again. 'Stations abroad have created commuter towns, work opportunities.'

'Resulting in house values rising,' says Councillor Cornell, estate agent, 'which can't be bad.' Discussions and arguments spool out like a loosed fishing line.

'Gentlemen, let's pause for tea.' Chairman reels them in. 'Wet our whistles.'

'At last,' says a parched voice. 'Ooh, biscuits; council being rash?'

Everyone takes their tea and duty seriously for another hour.

Councillor Sadler suggests, 'Perhaps we have been tardy in following the trend.'

'Or wisely let others test the water first?' Says another.

Chairman Merryfield: 'Is it time, belatedly, to grasp the nettle? At one-thirty a.m. it is certainly time to vote.'

'And return home *if* we can find a Hansom cab, at this hour?'

The vote passes, just: 9–7.

Now, two Regency seaside venues and busy villages sit starved of rail connections. Insufficient bus services, only, serve as lifelines to these picturesque places. One station is now a house; others were demolished and built upon. The line was promptly cleared of track, some land taken back. Residents and all those potential visitors have been driven onto the roads, literally, or other-ways deterred from travelling to these rail-impoverished pretty towns.

These days, two towns bravely put on annual music and literary festivals despite the convoluted routes and insufficient parking visitors must suffer to attend them.

All we're graciously allowed, on a small section, is the salvaged route recycled as a footpath and cycleway, courtesy of passionate volunteer groups tirelessly pressuring local authorities, to do the right thing.

The past can sometimes be repaired, with foresight?

Budleigh Salterton Station looking towards Tipton St Johns and Sidmouth Junction on the London and South Western Railway, now lost under housing (Ben Brooksbank)

SHRINE

GRAHAM POWELL

Placed across a downhill track
in the middle of the woods,
a damp and vaulted chapel

crafted out of tiered
and layered bricks
to brace the falling ground

the arches of the tunnel
are stepped and graded
with precise exactitude

to reflect the slope
and set a level track
for the railway

as it passes vacantly
through cuttings
and embankments overhead

the passage rests secure
there are no trains
to chunter down the line

the perfect form's redundant
an open nave
for the framing of the woods

fashioned at the corners
with coping slabs
of resilient railway blue

inside the air is damp
with space enough
to support

an echo
some glossy ferns
an idle trail of liverwort

a chapel left for walkers
as they pause
upon the upslope

to marvel at the trees

ARE YOU CARRYING A MISSILE?

ASMA KHAN

Do you have a missile in your bag?" The border control officer, a middle aged woman, smiled at me warmly. I knew it was a trap. She wanted me to share a joke with her, and once I jokingly said "Yes, I don't leave home without one", that would be the moment I would be hauled off the train for further interrogation.

I had booked a trip to the Balkans where I would travel on my own starting at Greece to see family, then taking a coach to Bulgaria, a train to Romania and then on to Budapest, and finally a hydrofoil to Vienna.

Unfortunately a week before my trip, the awful tragedy of 9/11 happened. My own relatives in New York had been impacted by the events, one witnessing what had happened. The repercussions were to have an impact on me for years to come: suddenly being an Asian meant I was met with hostility and suspicion in certain places, I was pulled aside for extra security checks at airports. My first experience of this was my flight to Greece where I had to unpack my hold luggage so every item could be checked. I understood the staff were just doing their jobs. I understood the suspicion, as hurtful as it was.

I was left largely alone on the coach to Bulgaria. It was on the train to Romania where I was asked if I had a missile in my backpack. I had smiled at the lady politely and said a firm, "No." She then left me alone.

The taxi driver taking me to a friend's house in Bucharest seemed friendly enough, until he started pointing out bullet holes in buildings, locking the taxi doors and demanding an extortionate fare. I was glad to leave Bucharest after my short stay there. Unfortunately at the train station I was bewildered by the train departure displays until a young man asked in English if I needed help. I explained I was looking for the train to Budapest, and he very kindly picked up my backpack and told me to follow him as he led me to my train. I thanked him profusely and gave him what few notes I had left, not much money at all, in that currency. He went away and I tried working out where my seat reservation was, but within a minute he had returned with a police officer.

"This officer would like to see your passport," he explained.

I handed over my passport to the officer. There was an exchange between the two in Romanian. Then the young man interpreted for the officer: "He says you don't have a visa to be in Romania."

I was suddenly on guard. I was wise enough to know this was another deceitful ploy. "I don't need a visa, I have a British Passport. My passport was stamped at the entry to the country." I indicated where the stamp was in the passport.

There was another exchange in Romanian. The young man said, "He says that was the entrance visa. You need a different exit visa and you need to come off the train and follow this officer to pay for the exit visa."

I had had enough by this point and snatched the passport out of the officer's hand as I knew that by keeping hold of my passport, he would have a hold over me. I had seen it done before on a trip through the former Yugoslavia.

"I don't need a visa and I'm going to stay on this train."

"If you don't come with the officer, he will follow you on this train to Budapest."

"Fine!" I said exasperated, "Let him follow me then!" And with that I went to find my seat.

Whatever bravado I had shown was just a façade. I felt shaken as I sat in my seat and worried the officer would indeed follow me. There was a nice girl sitting opposite me who noticed I was a foreigner and asked where I had travelled from. I told her about my experience with the young man and police officer.

"Oh we still have a lot of corruption in the country," she explained. "They have now set up a special department to try and stop the corruption in the police force."

Well that was progress, I thought.

Just as I finally got some sleep, we were stopped at the border to Hungary at 2am. My passport was examined and I was pulled off the train with my belongings. I was interrogated by a couple of officers, starting off with, "Do you have a missile in your bag?" (This was not a statement restricted to European border control, I was asked the same thing at customs when I flew to America a couple of years later).

I was made to unpack everything. The officers had an exchange over my English currency and one of them walked off with the £30 I had in English money. His colleague sniggered and said his colleague had gone to photocopy the notes as he collected foreign currency. In my naivete, I believed him.

The officer continued to quiz me, but at this point I did become annoyed at the suspicion. I felt I was a loyal British subject and being judged by appearance only. It was becoming tiresome and offensive. It was dangerous to be irritated, but he told me to pack my bags again and then left me alone for a while. Finally he and his colleague told me to follow them. My £30 was returned to me and I was put on another train to Budapest after 4am.

Thankfully, my journey was more peaceful after that but the experiences marred the last part of my trip.

When I returned home, I recounted my adventures to a friend. "I don't understand why the officer wanted to photocopy my English money though," I had commented.

"Oh he didn't photocopy them, he was waiting for you to offer the money as bribery."

Evidently I still had a lot to learn!

EACH DAY DOGS LEAVE HOME WITH NO KNOWING OF WHAT IS TO COME

BRONWEN R. EVANS

Back in London he'd get the train all the time,
short adventures, sprawling on the floor or sitting up
    while nodding off
held tightly on my lap. He preferred it to the bus
barking every time the doors opened, asking to get off.
    Not in fear,
just simple rightful impatience. I was the scared one,
taking the love most precious to me out into the
    unpredictable world.

When he and I were both young things we used to
    get the tube,
more excited by trips to vibrant places, new people,
    new smells.

A few years ago we happened across a tiny countryside
        steam rail,
took it as a family in the gentle breeze and warm sunshine
just after seeing a hare and a stoat.
He did not stop smiling. Warm fur and creased eyes
and all of us there were happy. His perfect train,
        small like him,
heading off into the green unknown,
a view of trees and wildflowers through half open sides.
Everyone he loved in one tiny wooden space.

Old suitcases on a trolley (Julie Wiltshire)

SUITCASES

JULIE WILTSHIRE

Waiting patiently on aged paint peeled wooden trolleys,
For the arrival of the 11:10,
Faded leather suitcases, phantoms of hope,
Stacked one on top of the other,
Devour the station with expectations,
Of far-off destinations.

Frayed straps, battered bodies and broken brass clasps,
Hold their secrets hostage.
They await in the mocking drizzle,
Listening intently for the panting breath,
Of their hissing, swearing saviour,
As he snakes his way around the distant bend.

The monster, avoiding the occasional land slip,
Grinds to a halt, announcing plosive consonants,
At the annoyance of his late arrival.
The brakes squeal upon the metal tracks.
Another broth of smoke-filled breath bursts forth,
Leaving its cloud of antagonism in the drizzled air.

A porter saunters down the empty station,
In his gloomy well-worn faded uniform,
And takes control of the soiled trolley.
The closed mouthed suitcases silently sigh,
As amongst the motes of dust and the brief day's light,
They know their freedom is guaranteed.

HALIFAX TOWN 1960s

STUART SAMUEL

Even before the days of on-train announcements, the most unfortunate visually-impaired passenger could tell when they'd arrived at Halifax. Just open the door, and it hit you – that sweet, warm, comforting smell of toffee coming from Mackintosh's factory just adjacent the station.

We're home, people would say, as they walked down the platform to the wooden stairs leading to the bridge and the exit. They faced a climb up Horton Street with heavy suitcases in the days before they came with little wheels attached.

The original Manchester & Leeds Railway, then Lancashire & Yorkshire, ran through the Calder Valley to Normanton, then northwards to Leeds, courtesy of Midland metals – a long way round. The owners wanted a shorter route, so the navvies built a difficult branch from Sowerby Bridge over the impressive Copley Viaduct and into the narrow Hebble Valley, in which Halifax had originated, as witnessed by the siting of the ancient parish church. As the settlement grew, it expanded up the valley side, but railway still occupied the narrow Hebble valley. Hence as weary returning holiday-makers faced the climb with their heavy loads, mutterings of the old quip could be heard:

"Oh why didn't they build the station nearer the town?"

"Because they thought it better to build it nearer the railway."

My parents never owned a car, so we relied a lot on trains to get around. Trips to Leeds for shopping or Blackpool for holidays were always by train, so I became interested from an early age. But I was not a trainspotter – the class of the locomotive or the number of rivets in the tender held no fascination. Rather, my interest was in their operation, and the people who worked with them. Hence, at the age of 12 when I was considered old enough to take two bus journeys per day from home to the Technical School on my own, I was also allowed to go to the station and explore its goings-on, and to make the occasional journey. It was at this time I became friendly with the three persons mentioned by name below. I was also invited into East signal box, but unfortunately cannot remember the name of the kind signalman involved.

In those days, the station had the suffix 'Town' to distinguish it from stations at North Bridge and St Paul's, King Cross, both of which had closed before I became interested.

It had six platforms – 1, 2 and 4 being for 'up' trains and 3, 5 and 6 for 'down'. In railway parlance *up* signifies lines to London and *down* those from. And in the case of west-east routes, like the Lancashire & Yorkshire, Manchester replaced London in prominence, as that was where its headquarters were situated at the city's Victoria station. It was a 'joint' station, owned by the Lancashire & Yorkshire Railway and the Great Northern, which had its own line from platforms 5 and 6 to Bradford via Queensbury via the then closed North Bridge – known affectionately by engine crews as 'over the Alps' because of its severe gradients and hard work involved.

Parts of it were still lit by gaslight, and on a cold, foggy November night, sitting huddled round the open fire in the waiting room on platform 3, the unrelenting passage of time marked by the slow sonorous ticking of the clock, wondering if the train was ever going to come is something I still remember after 60 years. And then the trill of a bell operated by the signalman warning staff that it was finally in section, and when it did arrive, it was either steam-hauled or a new, green diesel multiple-unit. For these were transitionary times, steam giving way to diesel on the hourly routes to Bradford, Leeds, Huddersfield, Manchester and Liverpool.

It boasted two signal boxes, Halifax East, and, guess what – Halifax West. East box was at the end of platform 3 and controlled the line to Bradford and West box at the other end of the station the entrance to the goods yard and the line to Manchester. The service pattern was hourly to Manchester, Leeds and Huddersfield, but for the hardworking signalmen there were three trains which were off-sequence. The first, a diesel unit, arrived from Manchester and instead of going on to Leeds, terminated and formed the 8:10am to York, leaving the way it had come in. It arrived on platform 3, and then having unloaded its passengers, went forward empty past Halifax East box to the crossover to the *up* platform 2, from which it then departed to York. That was a simple signalled move. But if it was late arriving, something the signalman at West Box dreaded on a wet morning, it unloaded and loaded in platform 3 and departed 'wrong' line to Halifax West, where it was crossed-over to the *up* line. The signalman's dread came because was this crossover was not passed for use by passenger trains because it was not lockable, so he had to leave the warm shelter of his box and clamp the points using a large G clamp under the blades to ensure safe passage of the train, and then display a yellow flag to the driver to inform him it was safe to cross. He then had to remove the clamp to return the points to normal. Meanwhile the Pennine rain used to blow horizontally from the west and he could get quite soaked in the time it took to do this.

A little later, a second one-train-per-day service arrived at the west end from Normanton, and again, unusually terminated at Halifax. Because it was overtaken by following trains to Leeds and Bradford, it used platform 5, to allow the following to pass. It then left at 9:48am for Leeds, and was a train I liked to catch, for as it was off-sequence, I often had it to myself, and as it was 'round the back' it was sometimes forgotten by the station staff. So come 9:48 the engines were running, the driver ready and the guard poised to press the buzzer, but no member of staff on the platform to give the right-away. The guard then had to go over the steps, cross to platform 3 and fetch someone out of the staffroom to come and wave the train off! Then the exit from the station could be difficult as platform 5 was at a slightly lower-level, and only seeing one train a day, the line could be slippery if it had been murky or drizzly for the last 24 hours, and the wheels raced, slipped and slithered up the grade out of the station as the driver gingerly made his exit.

The third one-off train was the most important for a young boy on his adventures, even though it comprised of only three coaches pulled by a steam tank engine. It left just after 9am and eventually ended up in London Kings Cross about 2pm. The three coaches were hauled to Huddersfield, where the tank engine ran round and took it to Wakefield Kirkgate. There the engine ran round again and pulled it up the incline to Wakefield Westgate and into a siding at the north end of the station. It waited for the main Leeds-London train and then propelled the coaches onto the rear, before the tank engine detached for a rest.

The main train now with Halifax portion tacked on eventually left Wakefield about 10:40am. So 1 hour and 40 mins into the journey, and I had travelled all of 20 miles or so to Wakefield! But it was appreciated by those who did not like changing trains. And it was so important, it warranted the Station Master, complete with uniform hat with scrambled egg (gold braid) on the platform to oversee its departure. Interestingly, that same person was under no compunction to greet the returning weary passengers around 11pm!

For me, and most passengers, on entering the station from the bridge over the goods yard and through the large double doors, the usual first port of call was the booking office.

This was situated across the wooden floor. From a small window, Alan, the booking office clerk, would sell travellers an Edmondson-type ticket (invented by Thomas Edmondson in the 1840s), made from pasteboard, approximately 1x2", pre-printed for popular destinations, but with a blank space for the clerk to fill in for the more exotic and less-frequently requested ones. Alan had an encyclopedic knowledge of the national railway system – he read railway timetables for pleasure, which was just as well, for in those pre-journey-planner days, devising an itinerary for a long journey involved wading through those timetable books, of which there were six to cover England, Wales and Scotland, corresponding to the railway regions. A request for two adults, one child and one dog return to Kingussie or Falmouth Docks would make his day – other clerks would be tempted to fob you off with local tickets to Leeds and request you book again there!

Then, clutching my treasured pieces of pasteboard I made my way to the ticket collectors at the head of the stairs leading to the platforms. Here, Kenneth Woodhead or Albert Clarke would greet me and clip my ticket. Each station had its distinctive mark – Halifax's was like a small Greek letter 'rho': ρ. They also collected the tickets from arriving passengers, checking they were all in order, and issuing a written excess ticket from a pad if necessary. Collected tickets were cancelled, bundled up in newspaper and sent each week to Newcastle. Why Newcastle? I like to think there was a clearing house there with an important group of auditors with nothing else to do but eagerly await each offering from Halifax, but, perhaps not. Maybe they just threw them into the Tyne. Another of Albert and Kenneth's duties involved turning the rollers of the big display cabinet showing the departure times and destinations of trains for the next few hours. Oh, and their most important job – bringing milk for the station cat, who'd quickly discovered the cosiest part of the station was in front of the small gas fire in the collector's bothy.

These, then, are my memories of Halifax station in the 1960s. It's still there now, modernised and reduced to just two platforms, but with a more frequent service.

- *Mackintosh's is now Nestle. (modernised with no escaping sweet aromas);*
- *The signal boxes have gone – it's now controlled electronically from York, not by signalmen but signallers, who can't even see the trains but just have a bank of computer screens to contemplate;*
- *The goods yard now has a children's museum on the site;*
- *The ticket office is rebuilt, along with self-service machines, but the collectors and bothy have gone;*
- *The Parish Church is still there, but now renamed Halifax Minster (not originally a Minster like York or Westminster, but I suppose it gives it more kudos on the tourist circuit – cynical? Moi?);*
- *One thing has remained constant over the years though – the climb up Horton Street …*

The goods yard used to stand in front of the Halifax station building, which was designed by Thomas Butterworth and opened in 1855 (Tim Green)

TRAIN TRACKS

MYLES CUTLER

Her hands wearing gloves in
the washing up suds.
He's asleep
in front of a repeat.
Midnight Train to Georgia flows from the radio.
She leans and dreams.

Ted reads a lot about
having a pension pot.
But all his nouse has
gone into his house.
Sell. What the hell.
This is his *Last Train to Clarksville.*

Her Dad has gone
but his records are left.
Lonnie Donegan
sounding mighty fine on *Rock Island Line.*
Dad's youth of maximum
fun on minimum money.
Homemade instruments.
He never even
took a flight.
Didn't want to.
Didn't need to.
It was all here.

Dave stands in the
stands. Sings
Time Flies by When You're Riding on a Train …
with the twelve thousand home crowd.
His little daughter on his
shoulders
shouts the *'Arghhh!'* on the last line.
Dave has never been happier.

Her boyfriend, her first real boyfriend,
offered a job, his first real job,
abroad.
"It's the job or me!" her shout.
He walks out.
A song on the player
'Good times gonna come,
World as one'
She calls, he answers
instantly.
Missed his connection,
deliberately.
He's coming back.
"No. Let's go forward.
I'm coming too."
Peace Train.

A poster for the Simplon Orient Express
(CIWL archives, Wagons-Lits Diffusion)

THE ORIENT EXPRESS: A TRAIN JOURNEY WITH A DIFFERENCE

CAROLINE BRADLEY

Mention the Orient Express today and the reaction is one of surprise, of awe even.

"Oh, you've travelled on it, have you? How exciting!"

You can almost see the multiple £ signs, reflected in the speaker's eyes as they try to equate you with the glitz and glamour of luxury train travel, as epitomised in Agatha Christie's *Murder on the Orient Express.*

The exclamation masks a subtext that reads: *I didn't think you had that kind of money; it just doesn't seem your style of travel, somehow.*

Images of waiters in tuxedos and white gloves, of cut-glass champagne flutes, gilt-edged menus and lavish dishes served on silver platters flit between us.

"I have," you reply brightly. A long time ago, as a student backpacker with some friends. We hesitate to add 'on the hippy trail' as that has other connotations.

The "Oh I see" and raised eyebrows tell me this is a wise omission.

"So, what was it like?"

The years fall away. I travel back in time to my early twenties.

A platform at the Gare de Lyon, the year 1974, the start of an overland trip across Europe to Istanbul and the Bosphorus, the gateway to Asia, and beyond.

The barrier opens and there is a stampede, a scramble to secure a place on the famous train. The number of ticket holders far exceeds the availability of seats. Of the long line of coaches, only two, the de-luxe ones will travel seamlessly through to Istanbul. Lettered on the outside of these are *Paris Simplon Venice Orient Express* which encapsulate the notion of travel and romance. The occupants of the others will have to track down the train onto which the reserved coaches will be shunted in Milan, Venice and Belgrade.

With heavy rucksacks, we are jostled and pushed, swept along the platform in a flow of bodies. Some manage to peel away to the edge and clamber aboard, shoving suitcases, carpet bags, children and elderly relatives before them, up the steps to where they hover, blocking the doorways. Others filter to the far end and work their way backwards along the narrow corridor, peering into every compartment for vacant seats. There is a constant click of latches, the grating sound of sliding doors being opened tentatively and then slammed shut again. The words *occupé* and *pas de places* hang in the air, along with a waft of garlic, stale sweat and cigarette smoke. There is a sense of urgency, of mounting panic, as passengers force their way through the carriages.

Jamming our rucksacks into a sliding doorframe, we stake our claim to two seats between the four of us in a compartment already occupied by six people. We draw lots for the first two-hour seated slot. There is a lot of shuffling and shifting. Two of our party squeeze in as best they can, the other two take up position in the corridor, perched on their rucksacks.

The sights of Paris give way to sprawling suburbs, distant hills and the wheat fields of the Paris basin. Our journey has begun.

Somehow, we survive the first night and wake skirting the shore of Lake Geneva. In Milan a system for changing trains evolves. We develop a *run-for-it-drop-it-scramble-and-bag-seats* routine. Unencumbered by rucksacks, two of us race ahead and secure seats while the other two follow, laden with luggage to set up a rucksack station in the corridor. Rarely do we manage to get more than two spaces. In Venice there is the inevitable push for the doors. People swarm, shout and swear, elbowing their way forward in the rush to queue for food, buy water and find the platform for the next waiting train. As for our little group, we survive on rolls, olives, bananas and Polo mints.

In Belgrade the scramble for seats comes to a head with passengers scurrying across the rail tracks to reach the train that will take us on the last leg to Istanbul. We follow their lead, hoping not to be mown down by any other incoming trains and struggle to keep the two through coaches in sight.

We find ourselves in a compartment with an Italian couple with a baby, a Muslim cleric, and several Turkish *Gast Arbeiter* taking annual leave from factories in Germany. They light up pungent Turkish cigarettes, which they offer around. Taking in my partner's beard, they call him 'Ali Baba', a name which sticks for the rest of the trip. This time we toss a coin for the first two-hour seated slot. Ali Baba and I lose out and perch on rucksacks, while the others ease themselves into the remaining space.

To cope with the extra passengers, two rucksacks, the toddler, food, and bedding rolls need to be stashed in the overhead luggage rack. In the process, the Italian couple's bundle of food unravels and a large garlic sausage, slab of goats' cheese, figs, olives and grapes fall out.

The lady gives a shriek of alarm as she tries to pick it all up and stop her child from stuffing whatever he can salvage from the floor into his mouth. There is a gasp of horror from the cleric at the sight of pork while the toddler begins to howl. The Turks, meanwhile, seize on the opportunity to divert attention from the plastic bottles of cheap red wine protruding from their own bags. In the confusion, the grapes are trampled on, the wine spilled, and the sausage split in two. At this point, the narrow corridor does not seem such an unattractive spot to be, after all. Not for long, however. Squirming with fury and indignation, the toddler is suddenly thrust out of the compartment into my arms. There he stays, rigid with shock, while his parents try to clear up the mess.

The cleric edges his way to the sliding door, and, gesticulating wildly, wedges himself in the door frame, adding to the chaos.

"Pig sassige, baby cry, bad drink and," pointing at my partner, "Ali Baba, hippy man."

Muttering, he brushes himself down and, with strips of sundried tomato and goat's cheese clinging doggedly to the hem of his robes, takes up position beside us in the corridor. The whistle blows. With a sudden jolt we are thrown tumbling back into the compartment. We are on the move again.

Bags are rearranged, the spilled food is scooped up and stowed with the offending sausage in the rack along with the toddler, now released from my arms. Wedged between two rucksacks, he whoops with excitement every time the train goes round a bend, his chubby legs dangle down, swinging from side to side until he falls asleep. The Turks play cards; the others chat or doze. All is peaceful as long as no one leaves their respective perches.

As dusk falls, everything changes: the cleric suddenly springs to his feet and, with age-defying agility, leaps onto the small table by the window. He reaches up and, removing the sleeping baby, deposits him on his mother's lap, before extracting a tin bowl from his carpet bag. Clambering over rucksacks and bodies, he makes his way to the toilet at the end of the carriage, where he is joined in his ablutions by four other clerics from adjoining compartments.

Later, back at his seat, he pauses before swinging himself and his prayer mat up into the luggage rack. Here, caught between bedding rolls and the roof, he tries to face Mecca and prostrate himself as well as he can, joining his fellow clerics in their evening prayers. *Allahu Akbar* resounds comfortingly through the compartment as the sun goes down.

Later, we try to get organised to get some sleep. This is not easy crouched in the corridor or inside with legs stretched across the aisle, resting on the knees of the person opposite. The Italian couple suspend their baby in a hammock across the aisle and swing him to and fro. With rucksacks laid out on the floor, under the sleeping child, we can make room for another person in the luggage rack.

Perching myself on Ali Baba's shoulders, I swing myself up and curl into position.

As the train wends its way across Eastern Europe, the outside speeds by to the clackety clack of steel on steel. The gentle motion has a mesmerising, almost hypnotic effect on us and the random group in our compartment 'bubble'. In Bulgaria we see storks nesting on a station roof, notice old men following oxen-driven ploughs. Women with drab linen headscarves sow seeds, hoe and pick produce by hand. Still in Communist hands, the country has, at this time, little mechanized agriculture. Near the border Communist police board the train and check all the carriages. They say little but there is a latent aggression in their manner. Their very presence casts a menacing air on this stage of the journey. It upsets the fragile equilibrium in each compartment.

We chat, we read, we doze, we daydream, we battle our way across the continent, determined no shred of reality will shatter our dreams of romantic adventure. Bleary-eyed and dishevelled, we emerge 72 hours later into Istanbul's Sirkeci Station, '*where East meets West on the rails*'. Surrounded by features of European Orientalist design, we marvel at the marble facades, granite bricks, and tilework on the walls. Sunlight illuminates the stained glass in the high, pointed-arch windows and we are reminded once again of a bygone era of rail travel. We realise we have covered over 3000 km. Our journey has not been in the iconic style which the name *Orient Express* and the grandeur of this terminal might suggest, it is true. However, it has left us with a store of memories: of people places and strange happenings and with a sense of having 'travelled' in the true sense of the word.

READERS IN CLOSE RANGE

REBECCA MCDOWALL

The constant clatter, a manmade steel lullaby,
A gentle sway of carriages as changing landscapes rattle by.

Flurries of greens and browns interrupted by
evermoving streams,
Glassey eyed travellers lost in simple daydreams.

Sitting by the window as the country hurtles past,
My thoughts much like the train rushing here
and there fast.

We pull in at a station, one I don't know
As you stepped aboard, a silhouette, boarding all alone.

You sat in the fading light, the table our only divide.
Pulling out a book, I felt something shift deep inside.

Not the sudden shock or spark like stories often tell
But a quiet tension one I couldn't quite quell.

The train thundered on in the evening summer haze,
As the pages you thumbed held your enthralled gaze

A flicker of a smile danced across your freckled face,
The pages transporting you to an entirely different place,

A silent bewitchment, as the miles hurried past,
Our eyes met only briefly, our chosen books a contrast,

No words were spoken, simply a silent exchange
Just the thrill of possibility between readers in close range

A feeling of comfort enveloped where we sat
A shared understanding, no need for polite chit chat.

As you returned to your romance, your mind far away.
I glanced back to Tolkien, his pages laid splay

In the evening sun the landscape began to soften and blend
And our shared journey seemed to transcend

We two strange souls sat opposite each other,
Yet the readers within us could recognise one another

The divide of the table soon became a shared space
Where our two separate journeys had become interlaced

I found myself drawn to the faces you made,
As your lips quirked up at your novel's escapades.

A captivating portrait of a reader mesmerised,
A journey in peace, of bookworms harmonised.

I turned the page, the audible whisper of the sheet,
An almost silent crinkle, small but complete.

The crinkle a catalyst at the close of the day,
Conversation flowing over the typed causeway.

The constant clatter of the wheels faded to the black,
As you shared with me your favourite worn paperback.

We exchanged our journeys, the places we'd been,
The books that had accompanied us along with caffeine.

In the midst of the shadows brought by the fast
approaching night,
The potential of friendship blossomed in the trainlight.

We laughed in tandem at the quirks of our reads,
Fantasy and romance, a reader's prayer beads.

A comfortable silence between us gathered once more,
As the call of the books gnawed at our core.

Within our separate worlds our passions became one,
The hours had long dissolved with the setting of the
summer sun.

Your final stop grew closer and brought melancholy dread,
A fear that this moment would be lost and left unsaid.

The rarity of our meeting left unacknowledged
on the breeze,
Not unlike the fleeting impressions of past rustling trees.

Yet upon your face it was written crystal clear,
That you understood the feeling I held onto dear.

The train began slowing, an unwanted descent,
I jumped into action not wishing to resent.

A scribbled down number hastily scrawled,
As the wheels began to sputter and stalled.

The doors were thrown open letting in the cold night,
As you stepped onto the platform and disappeared
out of sight.

Only the promise of coffee and books remained,
The flicker of a friendship that would be sustained.

Between London and the North many miles there may lay,
But these two readers would meet on the tracks again
someday.

MY TRANSPORTS OF DELIGHT

DIANA ALEXANDER

Quick! Quick! Come and look." My dad grabbed my suitcase and, giving me what I felt was a rather perfunctory kiss considering I had been away at university for the last three months, propelled me up the station platform to the top of the train. It wasn't like Dad to move at this speed, but he was obviously highly excited.

Something was missing. Where was the engine and where was the steam? And what was this strange contraption where the engine should be? "It's a diesel engine," said Dad, "this is the railway transport of the future."

I felt a bit sad and nostalgic for the 'puffing Billies' of the past but there was no doubt that my journey had been fast and smooth and, for once, I wasn't covered in smuts from when I'd stuck my head out of the carriage window.

I've always loved train travel – for a child who suffered from chronic car sickness the train was, for me, a real transport of delight. No wonder, when they felt I was old enough (about

10 …), my parents were happy for me to travel with my pals to boarding school, first to Wales and then to Yorkshire by train. I can never remember feeling the least bit queasy on the train – in spite of the smuts and smoke.

How fortunate it was for me that my husband, Malcolm, also loved trains which meant that we travelled for hundreds of miles across Europe and the United States. Large travel agents were not happy to fix up special rail trips but we discovered a privately owned firm which specialised in dealing with eccentric travellers like us.

"Where is it this time?" they would ask – and they never let us down.

The excitement of arriving at St Pancras Station in the early morning, giving a cursory wave to the statue of John Betjeman, who loved trains as much as we did, and boarding the Eurostar for Paris, was so much more thrilling than hanging about for hours in soulless airport lounges. Also, when travelling by air you can never be sure what's happening to your baggage – will it be there on the carousel when you reach your destination or lost forever? On the train you can keep sight of your belongings at all times, and knowing you often have to carry them yourself encourages you to travel light.

Arriving at Gare du Nord increased the excitement – would we be able to cross Paris in time to catch our connection? (we always did but often it was close) and before long we'd be speeding across Europe, bound for Cologne or Strasbourg and on through Germany to Saltzburg or Vienna, the home of music. Another world, so different from the one in which we usually lived.

Or we would leave the Eurostar at Lille and catch the TGV, travelling south across Shakespeare's 'vasty fields of France' at a speed which we had to admit, couldn't be matched by our British trains, until we reached the bustling port of Marseilles. There we changed to the coastal train which took us (rather more slowly) along the Cote d'Azur to Menton not far from the Italian border, admiring the sea and the incredible boats moored in the bays.

"You must be mad," our Menton friends would remark. "You lose two days of your holiday."

"But it's part of our holiday," we would tell them and it certainly was.

On the train we could move about and chat to other passengers (we were both good at that) and so often found someone with whom we had a connection. We once got talking to a couple from Northern Ireland and discovered that the woman was the cousin of a girl with whom Malcolm had had a teenage holiday romance … The food was an enormous improvement from the unidentifiable offerings in plastic containers which most airlines provide and there was always a bar where we could quench our thirst.

Our trip to Russia to meet American friends who were shortly to return to the US did not leave us enough time for a train journey from London to Moscow. But we managed to squeeze in a trip which for Malcolm, a tremendous admirer of the Russian writers Tolstoy and Dostoyevsky – who seemed to spend a lot of their time on trains (Tolstoy died on a railway station) – was one of the highlights of his train travelling holidays. I found it quite exciting too.

We took the night train from Moscow to St Petersburg accompanied by the group with whom we had spent time in Moscow. A Canadian girl played her guitar and sang Russian folk songs, and rather a large amount of vodka was consumed, so when we all decided we should get a bit of sleep in order to enjoy the wonders of St Petersburg the next day we all remembered how we had been warned about the dangers of bandits and drug thieves breaking into travellers' sleepers and stealing their money and valuables.

Our American friends, who were not the sort to panic about anything and had spent a long time teaching in Russia, insisted that we took with us a wire coat hanger with which to 'lock' our sleeper door. But the sleeping compartments were for four people and our two companions were a middle weight boxer and fluent Russian speaker so we felt we should be pretty safe. Even so, I slept with my handbag under my pillow. Sadly, St Petersburg was the end of our train travel in Russia and after a few days, we flew home from another grimy airport which had nothing in common with the magical city it served.

By now we had really caught the rail travelling bug and we decided that, before we got too old we must sample the delights of overnight American trains. We had already done 'short' trips from one set of friends to another – but this was to be the Big One. We booked to travel from coast to coast on the San Francisco Zephyr starting from Chicago and sleeping for two nights on the train. This turned out to be one of the most exciting and enjoyable things we had ever done – and we finally realised why comparatively few Americans have passports. It would take several lifetimes to explore this vast land.

As the train sped along, Malcolm approvingly cast his farmer's eye over the fertile fields of Nabraska at the source of the great Misisippi and it seemed no time before we were climbing the Rocky Mountains, then on to Denver, Colorado where we stopped for a long break; not long enough, though, for one of the passengers who was on the platform phoning his wife when he saw the train leaving. Quick as a flash, he nipped out to the road, flagged down a sports car and was at the next stop before we were …

On the second night we drew into Salt Lake City and left

very early in the morning - which was a bit disappointing as we'd hoped to have a look at the cradle of Mormonism. But there were few disappointments – this was truly a trip to remember. Everything was done for the passengers' comfort, the food which was excellent was served in a very smart dining car by white coated waiters and at the end of every carriage there was always a pot of fresh coffee. There were parts of the train which had viewing platforms for when we were passing somewhere even more spectacular than the rest. The sleeping quarters could be described as 'snug' and it was something of a mystery as to how some of the larger passengers got into their bunks but we heard no complaints. Malcolm and I, both very bad sleepers, slept better than we could remember, lulled by the gentle motion of the train.

On the final day we passed the edge of the Nevada desert where the tumbleweed was rolling around as it does in the films and then along a glorious coast road which took us into San Francisco, a city of some of the best fun we ever had. Truly, truly an unforgettable experience and one which we would never have had if it were not our obsession with Transports of Delight.

VISITING HULL AS A TEENAGER IN THE 1960s

MARGARET ROYALL

When we were in our late teens my school friends and I sometimes made a day excursion to the city of Hull, which sits on the bank of the River Humber, opposite the small town of Barton-on-Humber. In those days the only means of getting to Hull was to take the train from Cleethorpes to New Holland*, from where we boarded a paddle-steamer ferry across to Hull. Paddle-steamers were used, as they were able to cope with the shallow, shifting sands of the Humber. The river is tidal and on several occasions when crossing at low tide the ferry became stuck on a sandbank. The crew would fetch long, sturdy poles and push them into the sand to try and dislodge the boat. On

* The railway company had built a pier at New Holland some 1,500 feet (460 m) in length allowing direct connection with the ferry service. New Holland Pier railway station opened on 1 March 1848 and service ceased in June 1984 when the newly built Humber Bridge was opened.

one occasion this did not work and the passengers had to wait until the tide turned and the rising water floated us off the bank. The two ferries we most encountered were the Wingfield Castle and the Tattershall Castle.

The train journey to the ferry terminal took us at a leisurely pace, stopping at country stations and halts, where we would wave to the man in the signal box and the guard in his uniform with his flag and whistle. In spring we would see in the fields alongside the track, lambs and calves playing rough and tumble with their siblings or feeding from their mothers, while in autumn we were delighted by the activities of the harvest: enormous combine harvesters droning up and down, golden sheaths of corn stacked in bales (referred to as corn stooks back then), and seagulls following the machinery, keen to dine on the pickings.

Lincolnshire has always been an agricultural county and the everyday activities of farm hands, casual labourers and horse riders would be there on display for us to chat about en route. It was our Lincolnshire, a peaceful backwater that time forgot. I miss those days.

Kingston-upon-Hull, usually known simply as Hull, was a city with a greater population than Grimsby. Inland from the thriving fishing docks there were large department stores, art galleries and much to occupy us. It was definitely the place to be. On arrival we rushed to eat in one of the famous fish and chip restaurants in the city centre. As in Grimsby and Cleethorpes, the best fish and chips could be eaten in Hull, cooked in beef dripping, which makes all the difference.

We loved the opportunity to sample the latest lipstick shades in the cosmetic department, perching on high stools to apply the trendy colours like TNT, a pale, chalky and rather unforgiving shade that was supposed to give you a Marilyn Monroe pout. Our parents would have screamed "Take it off!" but they were not there and we had the freedom to walk around Hull looking like ghastly clowns if we wanted to. It was a rite of passage, I guess.

We were generally exhausted by mid afternoon so indulged in a frothy coffee in the nearest Wimpy Bar. Then it was back to the ferry terminal and the journey home, tired but still excited.

The station on New Holland Pier in the early 1980s, shortly before the ferry service ended (Martin Addison)

BLACKIE'S BRIDGE

SALLY MILLS

We raced through the fields one thought in our mind,
never too sure if we'd make it in time,
always intrigued about what we might find,
eagerly waiting to see the tracks shine.
Engulfed in fear at where we were going
or was that simply due to not knowing,
we slid on our bottoms down to the tracks,
ensuring our bags were safe on our backs.
Looking for steam rising out of the trees,
floating straight up or drifting with the breeze,
we'd wait for the whistle at Blackie's Bridge,
the sound of the engine leaving the ridge.
Enveloped in a screen of smoke and steam,
our lips tightly sealed suppressing a scream,
faces flushed red with naïve excitement,
savouring the time that we had just spent
lost in a world which to us was unknown,
left with a secret we never took home.

THE SIGNAL

F. DAVID POTTER

The red light stopped the traffic –
me also, right opposite that sunken lane of memories,
they come tumbling towards me
pricking out long-forgotten forays
into that magic land of the trainspotter,
my heart reacts with a flutter
as if changing its position
looking up that lane, I see him looking back at me,

He stands alone
with pencil and pad waiting excitedly
for the next train, and then the teddy boy
who forced his hand into the ants nest
no tears, just a scream and the help
of a station porter placing him back
in the safety of the station platform.

I'm with him now, I can almost
smell the creosote rising from the sun-baked sleepers,
the deafening silence amplifying
the buzz of bees. Then the snap
of the semaphore signal, he looks up
searching with an increased pulse,

which I also feel, when the traffic lights show green,
and I have to move on.

Waiting for a signal at Alresford signal box on the Watercress Line (Lorna Brookes)

Inside Wolverhampton's Oxley steam Motive Power Depot with 48700, taken with a Kodak Brownie 127 (Jeff Nicholls)

OXLEY CATHEDRAL: A FIRST VISIT

1963

JEFF NICHOLLS

Oxley Cathedral is not to be found in any list of sacred buildings, but to a young and impressionable thirteen-year-old Black Country boy, Wolverhampton's Oxley steam Motive Power Depot is a place of almost religious worship. Reached by cycling along a rough canal towpath, alongside the oily green waters, a cinder track leads him steeply upwards towards this holy place. At the rather unpromising entrance stands a battered cast-iron 'Trespassers Will Be Prosecuted' sign, but he turns a blind eye to it and chains his bicycle to a nearby fence post with a mixture of trepidation and anticipation.

Widening out towards its summit, the cinder track is bordered by a primitive and battered low fence, made from old boiler tubes. On the other side of this crude barrier there is a steep and grassy embankment dropping down to the canal. Looking across from here, acres of industrial buildings can be seen; they are part of the former Great Western Railway's Stafford Road Works. Over the years, hundreds of steam locomotives have been built and repaired here, but the works are on the verge of closure. In the opposite direction, the view is an elevated one over Dunstall Park Racecourse. Not that our timid and nervous trainspotter, still in his short grey school trousers, has eyes for the horses and their colourfully-bedecked jockeys.

Arriving at the top of the cinder path, the first thing which catches his eye is a massive 2-10-0 waiting to leave the shed to haul a heavy goods train. Its driver looks quizzically at the young lad. The first number goes in his notebook and immediately after, and several hundred yards away, the massive shed building itself comes into sight. Before it can be reached, however, there is a forbidding patch of rough, open ground to cross with the constant fear of being spotted by the Shed Foreman leading to abrupt ejection or, even worse still, an encounter with the Railway Police.

Summoning up his courage, he walks towards his ultimate goal; to his right there are lines of engines waiting their turn to be serviced; fires are being raked through, tanks watered and coal dropped into waiting bunkers from the tall coaling stage, topped by a rusting water tank. These are the days when the machine that has served the country for a century-and-a-half is seldom, if ever, shown a cleaner's rag.

Excitedly, and with shaking hands, he scribbles the engine numbers in his grubby red Silvine notebook; the air is alive with drifting steam and acrid, sulphurous smoke, of hissing sounds, of the clanking of fire irons and the ringing of metal-on-metal. Daring to walk between these filthy, rusting but live creatures, whose days are numbered, there is plenty of opportunity for the careless and anxious visitor to trip over or into the great smoking piles of ash and clinker, or gash an unprotected leg on one of the many bent and rusty implements scattered randomly around the ground.

Trying to exude an air of confidence which betrays his bowel-churning inner turmoil, he exchanges nods and tentative smiles with shed staff and footplate men, any of whom might order him to leave, employing words his young ears may never have heard. Much to our timid trespasser's relief, they never do. A quick headlong sprint across the final open piece of ground, where he will be painfully visible to the Shed Foreman, and the relative 'safety' of the side of the main shed building is gained. Here too, there is plenty to catch him out – all kinds of tangled and twisted debris lurks in the grass; a stumble could be very painful.

Built in red brick in the early twentieth century, the shed building is huge: a whole 150 yards long and 60 yards wide, its sides punctured by dozens of tall, industrial-style windows, each made up of over fifty individual panes, which have probably never been cleaned. Many of these panes are cracked or missing. Frustratingly, the window sills are a couple of inches too high for a not-yet-fully-grown adolescent to peer through. A tall, wide doorway pierces the side of the building, a solitary track leading to the inner sanctum.

Venturing gingerly inside, inch by inch, a most rewarding sight greets him. Gathered around two big turntables are ranks of resting steam locomotives. Many of them are cold and silent, some of them with intricate-looking components removed and lying on the ground beside them. A few of these sleeping giants have men working on them, preparing them for their next task, the clanging of spanners loud and piercing in the gloom. Some have their smokebox doors open, allowing a glimpse of their soot-encrusted innards.

Looking upwards, the roof is supported by a mass of iron girders; a great wooden trough above each track makes a valiant attempt at clearing smoke and steam from the building, but the air is still thick with that unique smell of the steam locomotive and the ticking, hissing and gurgling of these machines at rest. Beneath his feet, the uneven brick floor is covered in rainbow pools of oil and grease, waiting to catch out the unwary and to leave their indelible stains upon his footwear.

Normally, steam locomotives would be seen from

platform level; here our nervous 'shed-basher' has to crane his neck upwards as they dwarf him. Their driving wheels are even taller than his still-growing body. The crowning glory amongst the locomotives in Oxley Cathedral is that wonder of wonders: a Great Western locomotive, bearing cast numberplates, with tarnished brass numerals, and chimneys with copper rims. Better still, one might be seen with brass nameplates. The first one he notes is called *Birtles Hall*; it leaves an impression which will last a lifetime.

Sneaking round the cavernous building, using the cover of each engine to avoid detection, his scruffy notebook slowly fills up with numbers. Most of the locomotives have probably been seen before, some of them many times, but there is always the chance of a 'foreigner', a machine which has somehow sneaked up from the south coast or crossed the Pennines from far-flung Yorkshire. Seeing such an exotic specimen will give a week's worth of bragging rights amongst his fellow enthusiasts in the school playground. Upon his return home, all of these hastily scribbled numbers will be entered neatly in a proper logbook; numbers of locomotives previously unseen will be underlined in his favourite colour in one of the famous Ian Allen ABC books. These are so precious that they will be guarded safely for the rest of his life; their yellowing pages hold a thousand priceless memories.

From time to time, a locomotive gives a shrill blast on

its whistle, alerting everyone to its imminent movement. Jets of steam issue deafeningly from its cylinder drain cocks; trickles of hot water dribble from its injectors. The great clouds of smoke and steam it emits help to hide the presence of our young trainspotter. Creaking and groaning like an arthritic old man, the machine will move onto one of the two turntables, the rumbling of the rotating platform adding another layer to the soundscape.

Once he has made his sneaky way round the main shed building and jotted down all the numbers, his next stop is the 'scrap line' on a patch of waste ground at the rear of the shed. Here, British Railways store engines which have steamed their last, their derelict corpses await their final journey to the scrapyard. Shorn of number and nameplates, these sad and rusting hulks will become a playground in the school holidays and many happy hours will be spent playing in their cabs, 'driving' them all over the country.

From the scrap line, he reluctantly makes his way back towards the main entrance, sneaking along the opposite side of the vast building from the first approach. Here, two old steam boilers with chimneys at least ten feet tall protrude from a small annex, their purpose is to dry sand for the shed's steam locomotives, or even the occasional intruding diesel which insists on lowering the tone of the place. Another dash takes him past the windows of the Foreman's office; a silent prayer is offered up that his attention is elsewhere.

From this point, it is all plain sailing. More redundant locomotives are to be found on the other side of the coaling stage, their numbers are duly noted. Some of them are old friends, never to be seen again; they are parked here, awaiting a tow to the scrap yard, perhaps to be reborn as cars, fridges or washing machines. Another hundred yards, a few more tracks to cross, the cinder path to descend, and he is home and dry. The pilgrimage is over. For today. Oxley's magnetic attraction will be irresistible right up to that awful day in Spring 1967 when it will close, but the memory of his first visit will always stay with him.

Outside Wolverhampton's Oxley steam Motive Power Depot with 75006, taken with a Kodak Brownie 127 (Jeff Nicholls)

If you loved this book,
please leave a review ...

The best place to leave a review is on Waterstones.com or on Amazon

Reviews make our authors happy and help us to share news of their work with more readers. Thank you for your support!

Did you know that this book has been ethically printed in the UK on FSC certified paper made from a mixture of fibre from responsibly managed woodlands and recycled material?

Crumps Barn Studio
crumpsbarnstudio.co.uk